# Using Natural Resources for Energy



Table of Contents

# Using Natural Resources for Energy

Reader

Core Knowledge®

**www.coreknowledge.org**



Printed in Canada

ISBN: 978-1-68380-534-2

# Renewable and Nonrenewable Resources

Chapter 1

Three hikers prepare dinner at a campsite. One boils pasta over the blue flame of a propane gas stove. Another fries fish over a wood fire. A third roasts vegetables in a solar oven over the course of the afternoon.

The wood and solar energy used for cooking are **renewable resources**. They will not be used up completely. While neither resource is available everywhere or all the time, both are renewable. The sun comes up every day. Trees can continually be planted and harvested.

The propane in a camping stove is a **nonrenewable resource**. It takes much longer than a human lifetime for propane and other fossil fuels to form.

**Big Question**

What is the difference between renewable and nonrenewable resources?

**Vocabulary**

**renewable resource, n.** a resource that can be restored more quickly than it is used up

**nonrenewable resource, n.** a resource that cannot be restored as quickly as it is used

**Propane is abundant now. But once the natural gas that is used to make propane is all used up, more will not form fast enough for it to be used as an energy resource.**

## Availability of Energy Resources Varies

> **Vocabulary**
>
> **fossil fuel, n.** a fuel formed from the fossilized remains of organisms

Natural gas is abundant. So are petroleum and coal, two other types of **fossil fuels**. Abundant does not mean renewable, however. What makes a resource renewable or nonrenewable is how quickly it is used compared to how long it takes to become available.

Fossil fuels develop over thousands or millions of years. Humans have been extracting fossil fuels from Earth's crust for centuries. But the rate at which we use them is much faster than the rate at which they form. That is why they are nonrenewable even though they are plentiful right now.

A resource can also be renewable but not necessarily abundant. For example, some locations do not get enough wind to generate electricity reliably for people's needs. Even sunlight can be too scarce to serve as a dependable energy source in some locations.

**In a sunny location, sunlight is both renewable and abundant. In Antarctica during the winter, the sun only dimly lights the sky for a few hours a day.**

## Some Energy Resources Are Renewable

Renewable resources that are used for energy include wood, wind, flowing water, sunlight, and heat from Earth's interior. In some cases, these energy resources are used in their original forms to apply energy or force to something. For example, sails on a sailboat harness wind to push the boat across the water. In other cases, the energy is transformed to a new form. A windmill, for example, converts the energy of moving air to mechanical energy that turns gears in machinery. These days, most energy resources are converted to **electricity**.

**Vocabulary**

**electricity, n.** a form of energy resulting from the flow of charged particles

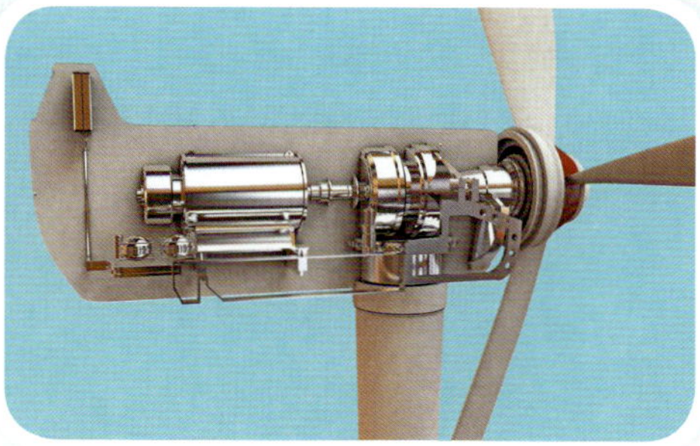

**Old Mills:** The windmill and water mill were both used to convert wind energy into mechanical energy to grind grain into flour.

**New Turbines:** This wind turbine and this water turbine convert wind and water energy into mechanical energy. A generator converts the mechanical energy into electricity.

## Renewable Resources Can Still Be Costly

The use of renewables is not necessarily problem free. For example, large wind turbines built to harness wind power can pose hazards to birds as well as aircraft. Renewable energy technology can also be very expensive. A homeowner who wants to turn his or her roof into a solar power generator may have to spend thousands of dollars to install solar panels.

Tapping renewable energy resources can also be harmful to landscapes, rivers, and other habitats. The dams that provide hydroelectric power in some locations may damage the environment around them. A dam may block fish such as salmon from traveling upstream to reproduce. The construction of a dam usually involves flooding a large area of land.

**The Three Gorges Dam produces an enormous amount of electrical energy. But it also displaced more than a million people.**

## Petroleum Is a Nonrenewable Resource

**Crude oil**, also known as petroleum, is a liquid formed over time by fossil remains that are under intense pressure. Petroleum is refined into many liquid fuels, including gasoline, diesel, and kerosene. Petroleum is also made into plastics, lip balm, and other products.

> **Vocabulary**
>
> **crude oil, n.** a liquid also known as *petroleum*

Petroleum is taken from Earth's crust by drilling holes deep into the rock. A large underground field of petroleum may be under enough pressure to force the petroleum out of the crust and up through a pipe. Petroleum can also be pumped from the ground. The hole through which the petroleum is taken is called a well. If a well is not controlled, it can blow out, causing an oil spill. Oil spills are dangerous to humans working on or near the oil well. They also can harm wildlife and ecosystems. These are some of the costs of using petroleum.

Petroleum can be transported by pipelines, trains, trucks, ships, and other forms of transportation. It is processed at a refinery, a place where crude oil can be changed into many different useful products. These are the benefits of using petroleum.

**Much petroleum is pumped out of the seafloor. Oil spills can occur when oil wells blow out, pipelines rupture, or oil tankers have accidents at sea.**

## Natural Gas and Coal Are Nonrenewable Resources

Natural gas is another fossil fuel. It often forms in the same type of location as petroleum, from similar remains of organisms. Like crude oil, gas can be extracted from Earth's crust and transported to other places. And like crude oil (petroleum), it must be refined before it can be used. Natural gas is used in many homes for cooking and heating. This requires a network of pipes to bring the gas to homes. Because gas is under pressure and flammable, accidents such as explosions can occur.

Coal is the third major type of fossil fuel. Unlike gas and petroleum, coal is a solid. It forms mainly from the fossil remains of plants that once lived in very wet habitats. Coal has been abundant and inexpensive for centuries. But mining it is dangerous. When coal is burned, it produces a lot of carbon dioxide and releases other pollutants into the atmosphere.

These fossil fuels are usually burned to transform chemical energy into thermal energy, heat. The transformation can cause the parts of a car engine to move or heat water inside an electric generator.

**A piece of coal takes millions of years to form but burns up in minutes.**

## Using Coal Has Costs and Benefits

Many power plants that supply electricity are powered by coal. In them, coal is burned below a large container of water called a boiler. Burning coal heats the water, which boils and forms steam. Because the boiler is sealed, the steam builds up. As the steam expands, it moves through pipes to an area of lower pressure. Here, the moving steam spins turbines. These turbines convert the energy of the steam into mechanical energy that moves a generator. This converts the energy of motion into electricity.

The sooty gas produced by the burning coal can be treated to reduce some of the pollution. In some locations, the carbon dioxide is directed underground. In general, however, most of the carbon pollution goes into the atmosphere. Air pollution from coal-fired power plants can also contain mercury, which can end up in the water. It accumulates in organisms that humans eat, such as tuna. These are some of the costs of using coal.

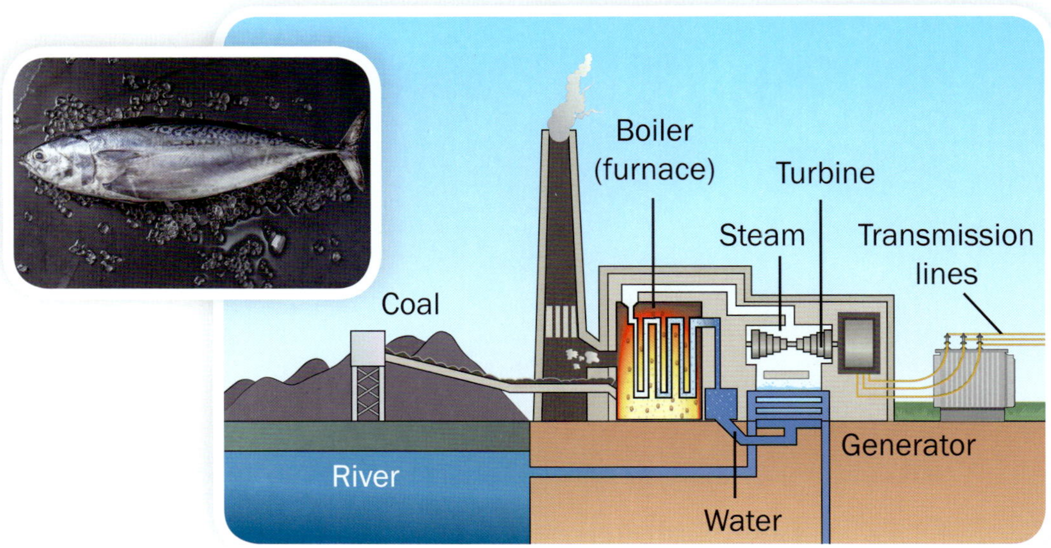


**Burning coal produces carbon dioxide and mercury pollution. Mercury can accumulate in marine organisms such as fish and then end up in humans.**

## Nuclear Power Is a Nonrenewable Resource

Another type of nonrenewable energy people use to produce electricity is nuclear energy. Nuclear energy has been used to make weapons, but it has also been used in nuclear power plants. In a nuclear power plant, the release of energy is controlled. Controlling the energy release allows heat to be directed into a boiler similar to what is used in a coal-powered plant. The boiler produces steam that spins a turbine, which makes a generator produce electricity.

Nuclear power does not release carbon dioxide into the environment, but it does have costs. If the release of energy is not controlled, the reactor can generate too much heat and melt part of the reactor. This exposes the surrounding area to radiation. Radioactivity is the release of harmful particles from a substance. Radioactivity can be deadly to organisms. It can also linger in an environment and cause a variety of diseases over a very long time.

In 1986, an accident occurred at a nuclear power plant near the Ukrainian city of Chernobyl. People for many miles around had to abandon the area to escape exposure to radiation. The area will not become safe again within the lifetimes of the people who had to leave.

# Types of Fossil Fuels

**Chapter 2**

A train loaded with coal rolls toward a power plant. The train is powered by diesel fuel. The moving parts of the train are lubricated with different kinds of oil. As the engineer controls the train, he uses a plastic fork to eat lunch from a plastic container. Later that day, he goes home and uses a butane lighter to start a small coal fire in a stove to heat his home.

**Big Question**

What are fossil fuels?

Every sentence in the paragraph above describes the use of at least one fossil fuel product. Coal, oil, and natural gas are three fossil fuels. People have developed many different substances from these fossil fuels, too.

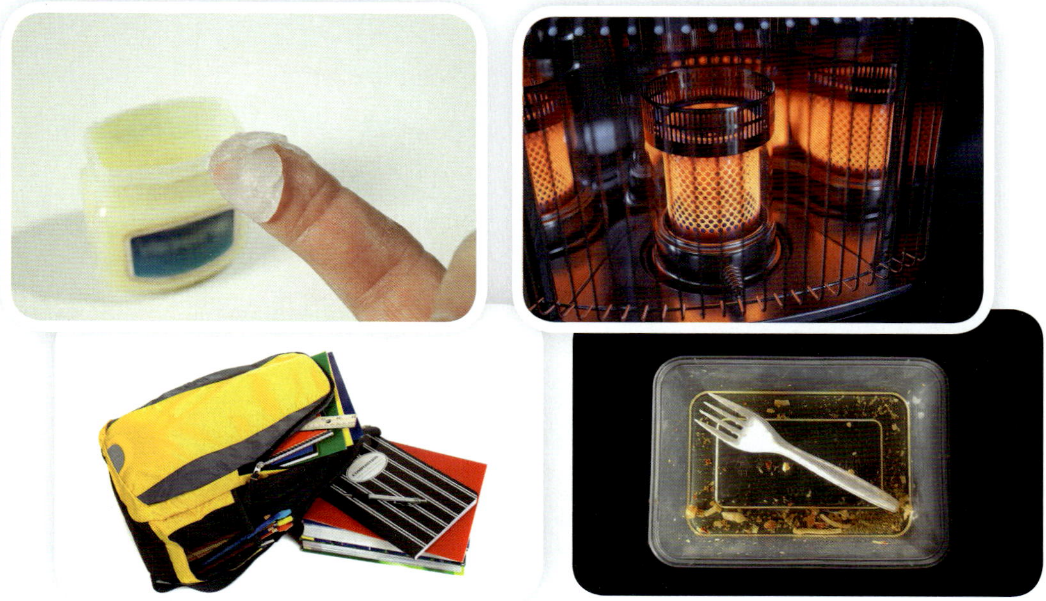

**(Clockwise from top left) Petroleum jelly is made from crude oil extracted from Earth's crust. Kerosene is a fuel made from petroleum that can light lamps and heat stoves. Plastics like this fork and container are made from oil. Nylon fabric in backpacks is made from oil.**

## How Does Petroleum Form?

Fossil fuels are made of the remains of organisms. We call them fossil fuels because humans use them to fuel processes such as heating homes, powering vehicles, and producing electricity. All three fossil fuels are made of the remains of organisms that lived long ago.

Some marine organisms that died and settled on the ocean floor were then buried by layers of sand or mud. That sediment became rock over time. The weight of that rock on the organisms' remains provided pressure and intense heat that turned them into petroleum, or crude oil. Oil companies that look for and extract petroleum can study samples of the ocean floor. They use different technologies to identify areas that may hold oil deep under the surface. This work costs a lot of money. But there is a lot of money to be made back, because oil is widely used by people for many purposes.

**Petroleum and Natural Gas Formation**

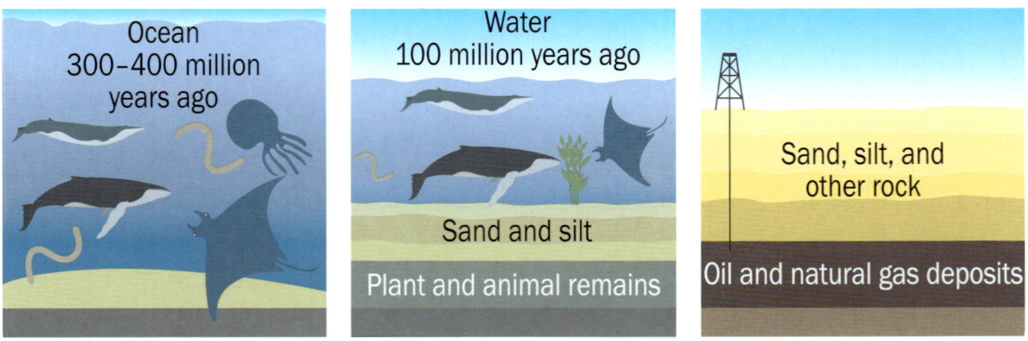


**Ocean organisms die. They become buried by silt and sand on the ocean floor.**

**The remains are buried deeper over millions of years. Heat and pressure turn the material into oil.**

**Petroleum deposits are reached by drilling through newer layers of rock.**

## How Do Coal and Natural Gas Form?

Coal forms in a similar way. Millions of years ago, Earth's landmasses were covered with plants. Shallow seas rose and fell, killing many of the plants. Some plants fell into swamps when they died. Their remains were buried under layers of mud. There was not enough oxygen available in the mud to allow the remains to decompose. Instead, they became a substance called peat. As more sediment built up over the peat, the peat pressed together and became solid. This is coal. Coal is very abundant and has been found all over Earth.

Natural gas forms where petroleum and coal form. Because gas is less dense than solids and liquids, it rises in Earth's crust until it gets stuck under a layer of rock. Sometimes natural gas is pumped out of an area above an oil field before the oil is removed. Methane is the most abundant gas in the mixture.

**This exposed rock on an Australian beach shows a distinct layer of coal. Pressure from the layers of rock above turned soft peat into this black solid.**

## Varieties of Coal, Petroleum, and Natural Gas

There are different forms of coal, petroleum, and natural gas. Coal that has been under the greatest pressure is harder, blacker, and shinier than other types. The lightest coal tends to be brown and is considered low grade. In between are forms that are black but fairly dull and not as dense as the shiny, high-grade coal.

Petroleum is a liquid in Earth's crust. The liquid can be very thin and light to thick and heavy. This means that some crude oil floats if spilled at sea and some sinks to the ocean floor.

Natural gas consists mostly of methane, but it can also contain propane and butane in gas form. Methane is piped into homes and businesses for heating and cooking. Propane is popular for use in small tanks for barbecues. Butane is often used in lighters.

**What begins as peat on the surface can eventually become coal. The deeper it is, the more pressure and heat it encounters. Different types of coal result from different amounts of pressure and heat.**

# Using Fossil Fuels

Chapter 3

People burn fossil fuels to release the energy they contain. We rely on fossil fuels to generate the energy we use in our everyday lives. Look under the hood of an automobile running on gasoline or diesel fuel. You will see and hear the moving parts and feel the heat rising from the engine.

**Big Question**

What are some costs and benefits of using fossil fuels?

There are many costs and benefits of burning so much fossil fuel on such a large scale. There are also costs and benefits in how these fuels are extracted from Earth's crust. And there are costs and benefits in how the resources are transported, processed, and transformed into usable energy. All this activity accounts for a large part of the world's economy.

It is hard to imagine life without these fuels. But our use of them is relatively recent. On the scale of Earth's history, using fossil fuels has had a major impact on the environment. Fossil fuels are nonrenewable. The costs of using them give us reasons to consider using renewable energy resources.

**This petroleum refinery produces many different products. Some will power vehicles. Others will become components of chairs, food containers, drinking straws, roads, and toys.**

## Extracting Petroleum

Petroleum is found only in some locations on Earth's surface. But most of what has not already been extracted is trapped under layers of rock. Some of this rock is on land, and some is on the ocean floor. After geologists identify an area that might contain petroleum, engineers use drilling technology to break through the rock and get it.

Before drilling through the rock, engineers must understand how much pressure the oil is under. They need to figure out how that pressure will be controlled. The weight of a thousand meters of seawater and many meters of rock on top of all that petroleum can cause oil to explode out of the crust unless the well is controlled. This means that all of the equipment that is going to extract, control, and gather the oil must be in place and ready to work before the extraction begins. These are some of the costs of drilling for oil.

**There are many costs to drilling for oil. Dangerous spills can release oil into the ground or large bodies of water.**

## Transporting Petroleum

Once an oil deposit is tapped, the oil needs to be collected and transported. If the rig is on land or close to shore, a pipeline can carry the oil to a refinery. But a spill from a pipeline can damage the environment. To some people, allowing a pipeline to be built is a chance to make money and add jobs to the local economy. This is a benefit. To others, a pipeline represents an environmental risk that is too great a cost.

Oil rigs that are far out to sea rely on oil tankers. These are ships that take oil from a rig and carry it elsewhere. But oil tankers can experience accidents, too. Oil tankers have run aground, capsized, sunk, and been captured by pirates. All these can add to the costs of transporting oil.

**This pipeline carries oil across desert terrain. The benefit is that it moves oil quickly from one place to another. If it fails, one cost is that it can do a lot of damage to the environment.**

## Refining Petroleum

Petroleum refinement separates petroleum into different forms of liquid oil. The different products can then be burned to generate mechanical, electrical, or thermal energy. Or, they can be turned into petroleum products, such as plastics and asphalt.

The process begins with the removal of oil from the ground. Wells pump oil from beneath land or offshore from beneath the ocean floor.

Next, the oil is heated. Different components of the oil boil at different temperatures. Vapors of propane, kerosene, and other parts of the oil enter a tall tower, rise, and then condense back into liquids at different parts of the tower. The different liquids are then collected and directed out of the tower to be stored or diverted to other parts of the refinery.

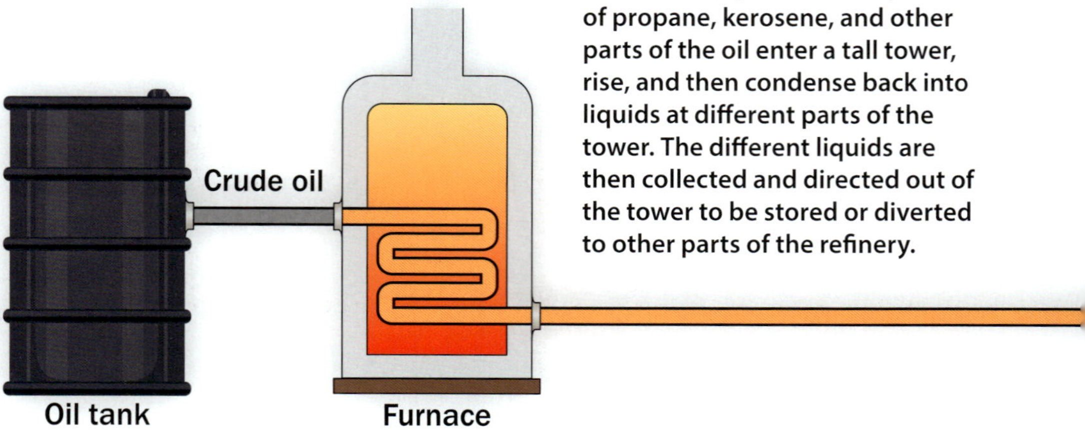

**Gases** and **thin oil** collect at the top.

**Liquid** and **thicker oil** collect in the middle.

**The stickiest oil** stays at the bottom.

**The separated liquids can also be transported to other locations to be processed into other products. For example, some of the components of the synthetic rubber in a basketball are produced by refining petroleum. Nylon, a plastic that is used in clothing and countless other products, is another product of the petrochemical industry.**

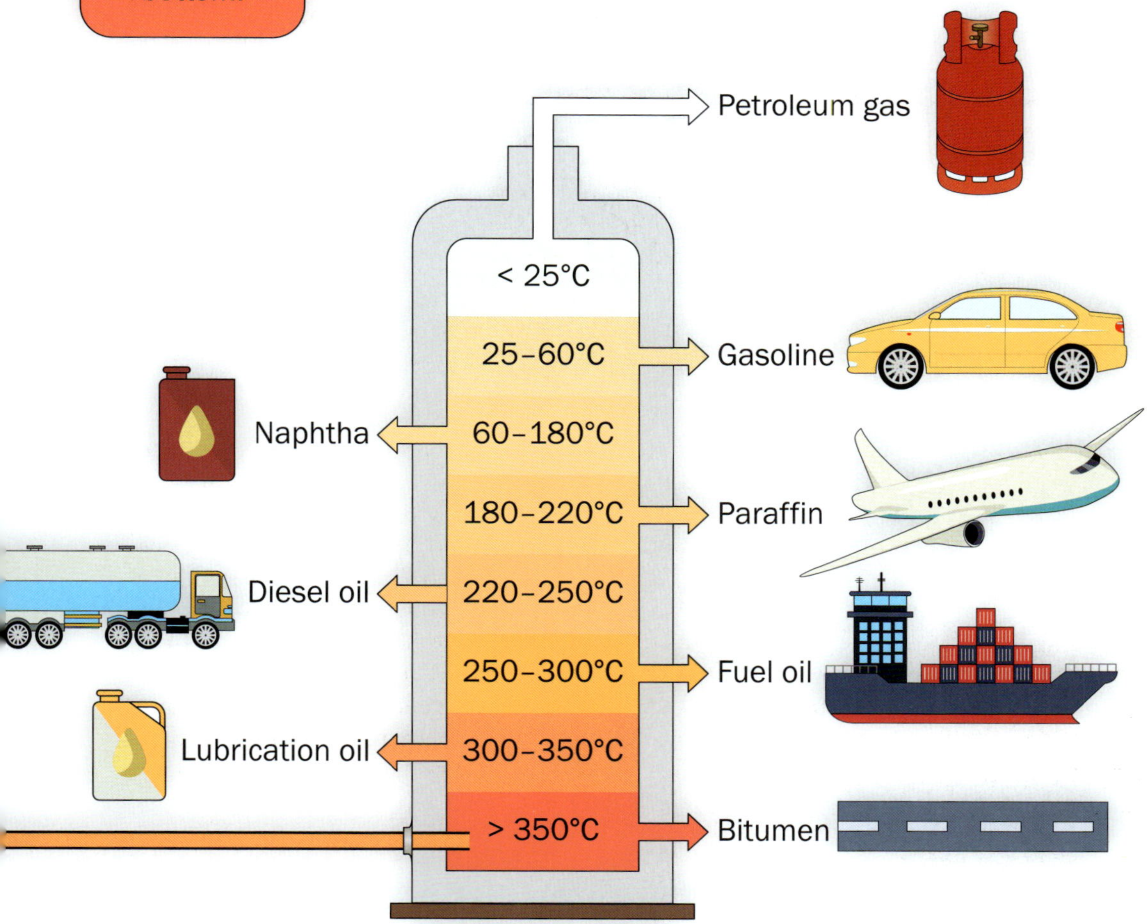


**Distillation tower**

## Coal Mining

Coal is a solid that must be mined. This means it must be dug out of the ground. Coal continues to provide heat for homes and cooking, as well as heating iron ore to produce metal objects. These are some of the benefits of coal.

There are many costs to coal mining as well. It is one of the most dangerous jobs. Mines can be filled with poisonous or explosive gases. They can also collapse, trapping miners. When washed or burned, coal can cause pollution. Coal can be transported by train, truck, or ship. These are less likely to result in an environmental disaster. However, some coal dust can escape into the environment if coal is not properly transported and stored.

**The photo shows West Virginia coal miners descending into a mine in 1938. They used shovels to load the car with coal by hand. These days, mining operations use more machinery, which reduces the physical stress on coal miners.**

## Extracting, Transporting, and Refining Natural Gas

Natural gas is often found where petroleum is found. It is common for the extraction of the two fuels to occur together. A drilling operation may first draw out the gas at the top and then proceed to remove the liquid below. Like petroleum, natural gas can be directed into a pipeline or onto a large ship to transport the gas for processing.

Large pipelines transport large quantities of gas across regions. Smaller networks of pipes carry gas to individual buildings. Many homes or businesses use natural gas for heat. This is one benefit of natural gas. Natural gas also pollutes less than other fossil fuels, but it still pollutes. Typically, gas-powered heating systems have underground gas lines connected to a main line that runs under a street. Burying the pipes helps protect them from being damaged. A broken pipe can become a fiery hazard if the gas is ignited. This can be both dangerous and costly.

**Cooling natural gas until it condenses into a liquid allows a large ship to carry about 600 times more methane than it could if the substance were still a gas.**

## Burning Fossil Fuels

Burning fossils fuels has many costs and benefits. Fossil fuels can be turned into many different products, but mostly they are burned for energy. They help heat our homes and cook our food. They also help power engines and provide electricity.

When fossil fuels are burned, the chemical parts react with oxygen in the air and become new substances. The chemical products of the reaction are mostly released into the atmosphere. This affects Earth's systems.

Burning any fossil fuel produces carbon dioxide. Carbon dioxide is a **greenhouse gas**. Greenhouse gases trap heat on Earth's surface. The more carbon dioxide there is in the atmosphere, the hotter the planet becomes overall. This is a serious cost of burning fossil fuels. Many scientists have stated that large-scale burning of fossil fuels is causing Earth to get warmer. Carbon dioxide that dissolves in water also makes water more acidic, which can make life difficult for animals such as corals and clams.

**Vocabulary**

**greenhouse gas, n.** a gas that traps heat in Earth's atmosphere

Greenhouse gases in the atmosphere are warming Earth's overall temperature. The thick ice of Earth's coldest places is melting as the climate warms.

# Nuclear Energy

Chapter 4

There are many types of energy that people can use to meet their energy needs. One of them is **nuclear energy**. Like other sources of energy, nuclear energy has costs and benefits.

**Big Question**

What is nuclear power?

About twenty percent of the electricity generated in the United States is provided by nuclear energy. That electricity can provide power to many homes, businesses, schools, and other places.

**Vocabulary**

**nuclear energy, n.** energy produced by the splitting of atoms, tiny particles of matter

## Materials for Releasing Nuclear Energy

Nuclear energy is generated when atoms of certain materials are split. This breaks the bonds holding the nucleus of the atom together, generating different atoms and releasing energy. The most common material to use for this process is uranium.

Less than one percent of the uranium found on Earth is the kind that can be used for nuclear energy. Uranium is found in rock usually mixed with other elements and even other kinds of uranium that cannot be used in nuclear reactions.

Uranium has to be mined and then enriched.

## Nuclear Power Plant

A specific type of uranium must be used to release nuclear energy. This material is known as uranium-235.

Useful nuclear energy is released from uranium-235 in a nuclear power plant. Look at the image closely to see how nuclear energy is transformed to electricity. The process called a chain reaction that releases energy from uranium-235 takes place in the reactor core. The fuel rods control the chain reaction. The containment shell helps prevent radioactivity from leaking into the environment outside. Radioactivity can kill or damage living organisms.

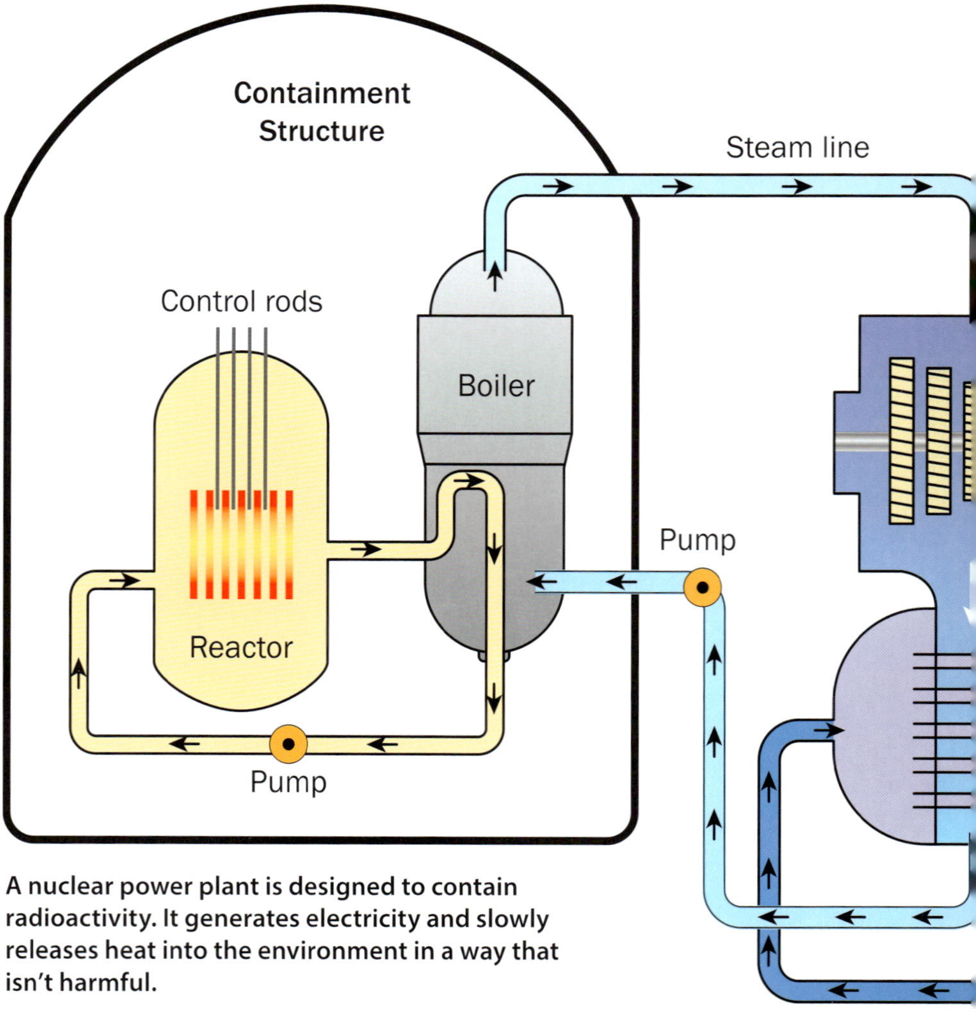


A nuclear power plant is designed to contain radioactivity. It generates electricity and slowly releases heat into the environment in a way that isn't harmful.

When atoms are split in a chain reaction, heat is generated. This heat converts liquid water into steam. The steam passes through a turbine, which powers a generator. The generator produces electricity, which is either stored or sent to power homes, businesses, and other structures. The steam then condenses back into liquid water. The water is cooled by pipes carrying water from outside the plant, usually from a river, lake, or sea.

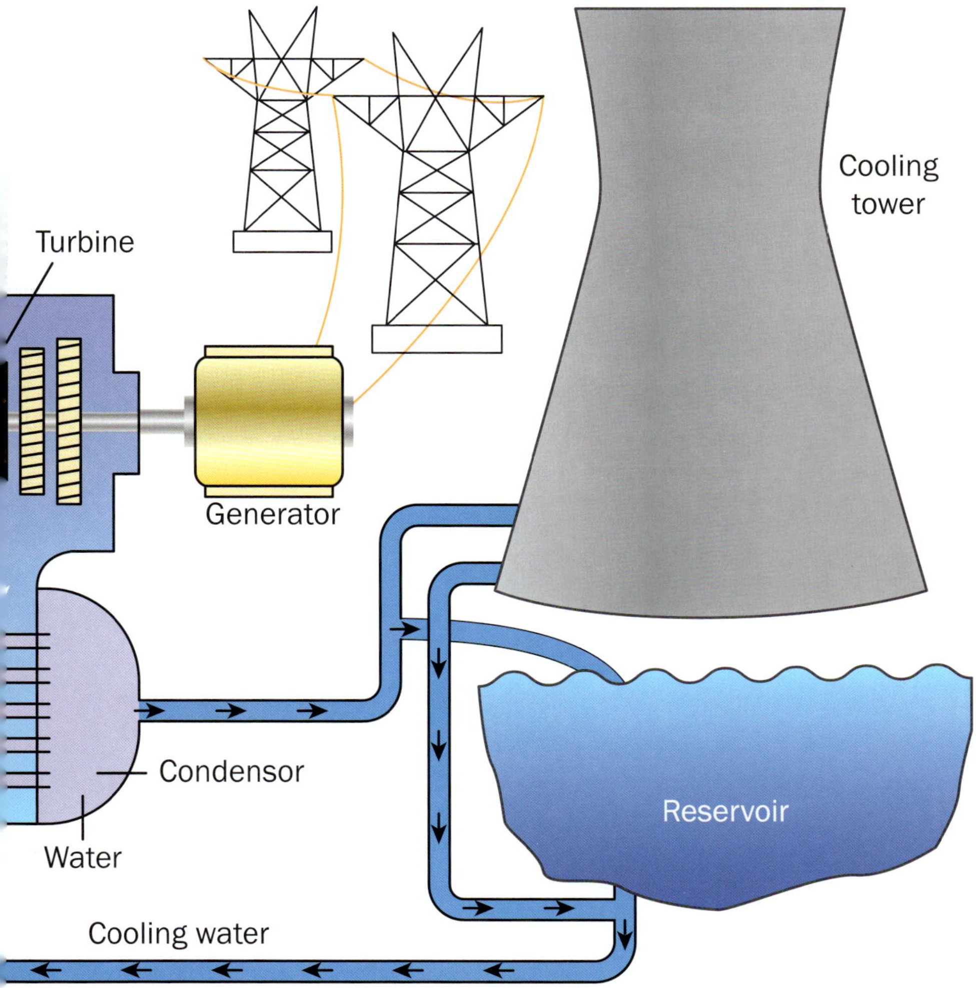

## The Costs of Nuclear Power

There are many costs and benefits in using nuclear energy. Some of the benefits were discussed on previous pages. One problem is that nuclear energy is nonrenewable. The materials that are the fuel for a nuclear reactor are rare. Also, nuclear power generates a lot of dangerous waste. Some of the material quickly breaks down into safe substances, but some remains radioactive for tens of thousands of years. This means that some materials need to be placed in sealed storage facilities that must be guarded or kept secret for centuries.

Nuclear material has been used to make weapons in addition to generating electricity. Nuclear bombs have devastating power. Because one nuclear weapon can do so much damage, governments have struggled to control access to nuclear materials and technology. A nation that claims to want nuclear technology only to provide electricity could also want it to develop weapons.

**Nuclear power offers a reliable way of generating electricity without producing a lot of air pollution. What you see coming out of the cooling towers pictured here is steam.**

# Wind Energy

Chapter 5

The sun heats the sand, trees, soil, and shallow waters around a tropical island. In the afternoon, the air over the island is very warm and rises higher into the atmosphere. This causes cooler air from over the sea to blow in. The moving air produces waves on the water. Boats that had been using their diesel engines are able to hoist their sails and use a free source of energy. This is wind energy.

**Big Question**

How do people use wind as a source of energy?

Near the island stands a series of tall wind turbines. The blades of the turbines rotate faster as the wind picks up. Inside each turbine, a generator produces electricity with each rotation of the blades. The electricity transfers through wires to a small power station. The station shuts down its noisy diesel generator. For the afternoon, the entire island is powered by the wind.

**Moving air, or wind, can move a sailboat. It can also spin a turbine, allowing wind energy to become electricity.**

## Converting Wind Energy to Electricity

For centuries, windmills converted wind energy to mechanical energy. The blades turned a shaft that was connected to gears. These gears helped operate a millstone for grinding grain into flour.

Modern wind turbines that convert wind energy into electricity have the added step of converting mechanical energy into electrical energy. In a modern wind turbine, the blades are attached to a large shaft with a large gear housed inside the turbine. The large

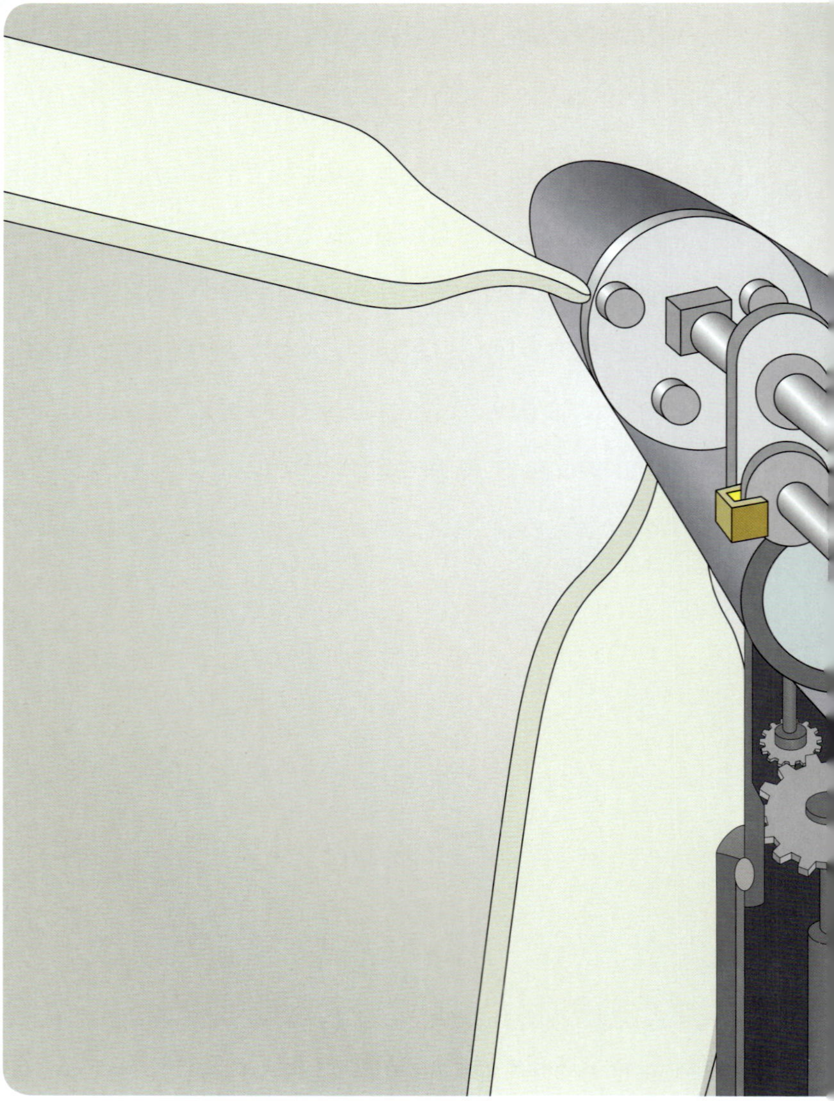

**The relatively slow speed of the large shaft in the turbine is converted to high speed to generate electricity. This technology is made of heavy parts, so turbine towers need to be very strong.**

gear rotates and is engaged with a smaller gear. The smaller gear is attached to another shaft inside a generator. The shaft rotates between 1,000 and 1,800 times per minute.

The direction that turbine blades face can be changed so the blades capture more or less wind. This allows engineers to prevent blades from spinning too fast when the wind is blowing too hard. Or it can prevent the blades from spinning at all when the wind is too weak to generate power.

## Things to Consider About Wind Power

Wind is an abundant renewable resource, but it is not always reliable. Some locations can be windy one day and calm the next. Some have winds that blow for weeks and then periods of calm. Others have breezes that are steady but not strong enough to turn large turbine blades fast enough to generate electricity.

A **wind farm** consists of multiple turbines placed together. Wind farms can cover vast areas. Turbines need to be spaced far enough apart that each

**Vocabulary**

**wind farm, n.**
many wind turbines installed together in an area to generate electricity for a community

one receives wind without disrupting how the others operate. Some wind power projects have been controversial because some people do not like the appearance of the turbines across the landscape.

Another factor that must be considered is how the turbines will be maintained and monitored. A turbine anchored to the ocean floor ten miles from the coast may be more difficult or expensive to maintain and monitor than a turbine on land. On the other hand, the offshore turbine may produce more power.

If you lived on the shore or even just went to the beach, would you like seeing this offshore **wind farm** on the water? Or would it bother you? If you were a fishing boat captain, would you appreciate seeing a renewable energy project in the waters you fish, or would you mostly be worried about having to navigate around it?

## Wind Energy Transmission and Storage

Wind energy harnessed by a turbine becomes electricity. Transmission, or the movement, of electricity usually involves power lines that run from a turbine to the nearest electrical power grid. The grid is the network of power lines, towers, underground cables, transformers, and other technology that distributes electricity to homes, schools, and businesses. When a wind farm generates electricity, the people who operate the grid can direct the electricity to specific parts of the grid and keep track of how much is being used.

**In the future, renewable energy use might look more commonly like this—homes with solar panels on their roofs and a wind farm nearby.**

Electricity can also be stored in batteries. Because wind power is dependent on wind, which is not always blowing, batteries can help make wind power more reliable. Individual homes or neighborhoods can even have their own small wind farms. A large battery can store the electricity for later use. Wind power can be combined with other electricity sources.

### Word to Know

A *power grid* is a network of power plants, transfer stations, and power lines that supplies electricity within a region.

# Hydroelectric Energy

Chapter 6

Like wind energy, water energy has been harnessed for thousands of years. It has always been used to produce mechanical energy, the energy of motion. But only recently in human history has moving water been used to produce electrical energy. **Hydroelectric power** is an important renewable energy resource.

**Big Question**

How do people use moving water as a source of energy?

**Vocabulary**

**hydroelectric power, n.** electricity generated by the energy in moving water, usually through a dam

Old water mills converted the energy in moving water into mechanical energy. A modern hydroelectric power turbine does the same thing.

Hydroelectric power is more predictable than wind power because water can be contained, or controlled. A hydroelectric power plant involves a dam that creates a large reservoir. The reservoir provides a steady supply of water for the turbines.

Hydroelectric power involves first positioning water above turbines. Gravity then pulls the water through the turbines.

## Things to Consider About Hydroelectric Power

A hydroelectric dam can be set up where there is a lot of water uphill from an area where it can flow down safely. It is most practical to set up a dam where flowing water can steadily refill a reservoir. This is why hydroelectric dams are usually built on large rivers.

Engineers must consider the source of the river water. The water cycle brings water onto land in the form of precipitation that falls when clouds cool and condense. When rain falls, it flows downhill and builds up in streams. The same occurs when snow and ice melt. But some of the flow from melting ice can be from precipitation that fell long ago and accumulated over a very long time. A glacier that is slowly melting and feeding water into a river

**This hydroelectric dam in Austria relies on seasonal snowfall that becomes meltwater, which fills the reservoirs. Without winter precipitation, the dam would be less productive.**

may eventually melt away completely. Then the river could shrink to a trickle or vanish completely. A more reliable river over the long term might be one that receives water from rainfall and snow that melts each spring and summer. Either way, engineers must consider data about the weather and climate of a given area.

Hydroelectric power depends on large amounts of water spinning turbines to generate enough electricity to make the project **cost effective**. Blocking a river so that a large reservoir will form behind the dam might be necessary to make the project cost effective. But this takes a lot of time, money, and careful planning.

**Vocabulary**

**cost–effective, adj.** describing the balance of factors when benefits outweigh costs

## Environmental Costs and Safety Hazards

While hydroelectric power is very clean, it is not without costs to the environment. Blocking a river to produce a large reservoir floods the area behind the dam. This destroys some habitats. Dams can also interfere with fish such as salmon. Salmon are born in rivers and migrate to the sea. Then they return to the streams where they hatched to reproduce. A dam also alters the river downstream. The flow of the river may be slowed and its depth reduced. Organisms that once had a fast-moving, wide, deep river might have something slower and narrower to swim through than before. Or the river may change dramatically from one week to the next as the dam controls the flow of water to meet the region's electrical needs.

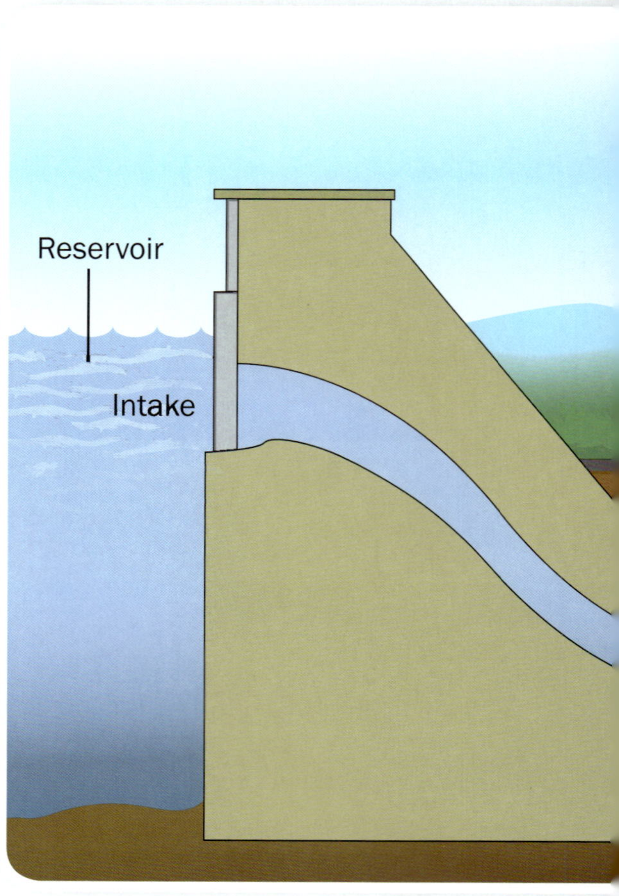


Hydroelectric dams generate electricity. Water moves from the reservoir into the intake. It spins a turbine, which is connected to a generator. The generator then generates electricity, which is sent through power lines to power homes and businesses.

Fish ladders can be built to help migrating salmon move up and down the dammed parts of the river. There are also newer dam and turbine designs that are "fish friendly." These allow fish to pass through with less interference. However, a dam's overall disturbance to river environments is hard to completely overcome.

Dams are also vulnerable to accidents. If a river or reservoir receives an unusually large amount of water from high rainfall, the pressure on its concrete and steel walls may be too great. This occurred in Laos in 2018, when a multibillion-dollar dam that was still under construction collapsed after heavy rains. A large amount of water suddenly spilled through the dam and flooded the landscape.

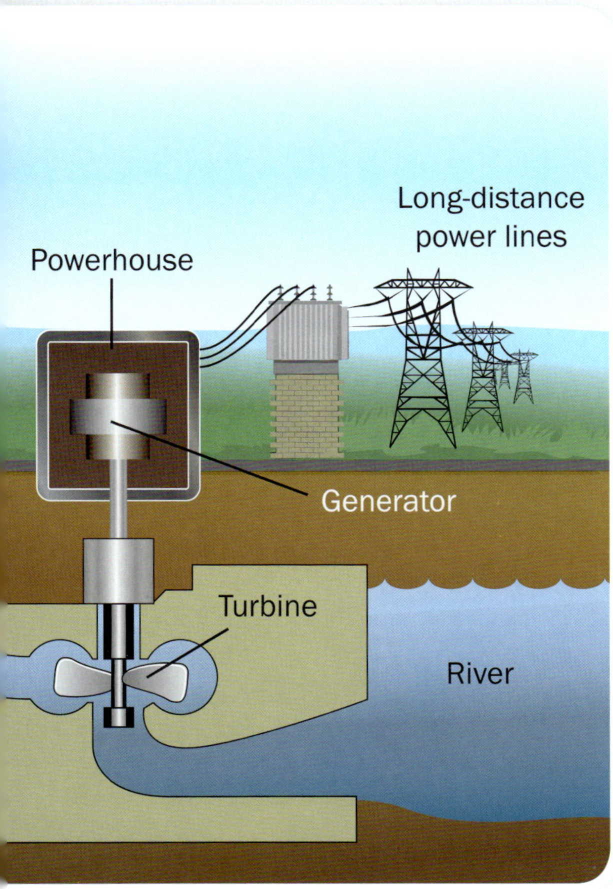

## Hydroelectric Power and Tides

Water moved downhill by gravity is one renewable resource that can be harnessed. Water moved by tides is another. Tides are caused by gravity, too. But instead of pulling water downhill, tides pull ocean water across Earth's surface. Tides result from forces of gravity between the moon and the sun and Earth's water.

To harness tidal power, turbines can be placed in areas where tides produce a strong current. A shallow span of ocean is often the best place because tides tend to move swiftly there. A dam-like structure can also be installed in an area with strong tides. As the tide comes in, the turbines spin one way. When the tide goes out hours later, the turbines spin the other way, producing electricity.

Hydroelectric power that uses tidal movement is practical only in some areas. Because tidal movement changes speed and can slow to a standstill, an electric grid is unlikely to rely on tidal hydroelectric power alone.

**Narrowing the area through which tides can flow makes transforming hydroelectric power into electricity more efficient.**

# Solar Energy

Chapter 7

Most energy on Earth begins with the sun. The sun transfers light and heat to the geosphere, atmosphere, hydrosphere, and biosphere. Unequal heating of land, water, and air results in wind energy. Evaporation keeps the water cycle going, which allows hydroelectric energy to be harnessed. Sunlight captured by the biosphere results in carbon-based foods that provide energy for life, as well as the raw material for fossil fuels.

**Big Question**

How do people use sunlight as a source of energy?

**Solar power** is the use of sunlight to meet energy needs. A **solar cell** is a technology that converts sunlight to electricity. Multiple solar cells can be assembled into panels, which can heat buildings or charge batteries.

**Vocabulary**

**solar power, n.** the use of sunlight to meet energy needs

**solar cell, n.** a device that converts sunlight to electricity

**Each of these shiny panels is a solar cell that converts sunlight into electricity.**

## Solar Energy's Costs and Benefits

Every hour, enough solar energy strikes Earth's surface to meet humanity's energy needs for an entire year. With that much renewable energy available, why aren't we using solar energy for everything? One challenge is the problem of solar cells. For decades, solar cells, also called photovoltaic cells, were inefficient. They could only convert about six percent of the energy in sunlight that struck them into electricity. Since then, efficiency has improved to about twenty percent. Higher-efficiency cells are made, and they are expensive. They are mostly used for satellites and other big projects.

**Word to Know**

How *efficient* something is refers to how quickly, easily, and well it does what it is intended to do. For example, if it takes more work and energy to make a solar cell than the energy and benefit it can provide, it is not efficient.

**A large solar farm can harness a lot of sunlight and provide a lot of electricity, but what happens when the sky is cloudy or the sun goes down?**

It doesn't make sense to spend a lot of money on solar cells that can only convert a little energy. But recent improvements and more demand for solar cells have brought costs down. It can still cost thousands of dollars to put enough solar cells on a home to provide electricity. For most people, that is too great an expense. And they do not want to wait years for that investment to pay off in lower electricity bills.

There is another challenge to solar power. Different amounts of sunlight strike Earth in different locations and at different times. The sun does not shine in the sky all the time. So, even if a home has solar panels or a town is using a solar farm to provide electricity, there still needs to be battery storage. Batteries capture the electricity for cloudy days and sunless nights. Batteries are costly, too.

## Indirect Use of Solar Energy

Solar energy can also be converted to electricity by another method. This involves reflecting many different rays of sunlight toward a single target that becomes superheated. The target is usually the top of a tall tower in the middle of a field of computer-controlled mirrors. Liquid inside the target becomes heated to generate steam. The steam powers a turbine, which produces electricity in a generator. Hot liquid can be stored in tanks and directed to the boiler later in the day when the sun is low.

The heat that is absorbed by sunlit materials can also be used to heat water for use in homes. Some homes are equipped with rooftop water heaters. A pump carries water or some other liquid to the rooftop unit, where it gets warmed by the sun. That liquid is then directed into a tank inside the home. Water in the tank is warmed by the heat from the liquid inside pipes that came from the roof. It is available for use in the kitchen sink, shower, and other places.

Some people also use solar ovens to cook food. The oven usually consists of a clear-roofed container that has mirrored walls. The sunlight that enters the container reflects off the walls and strikes the food inside. A solar oven usually cannot achieve the temperatures of an electric, gas, or coal-fired oven. But if enough time is given, it can cook some food very well. And the sunlight is free!

**This solar thermal energy plant in California uses thousands of mirrors to reflect sunlight at a tower. The top of the tower has molten salt inside. The salt is heated and directed into a boiler, which produces steam that drives a turbine and generator.**

## Passive Solar Energy

Another way for people to use solar energy is to design buildings so that sunlight warms them inside. For example, a home that has a large living room that faces the sun can have large windows that allow sunlight in. The floor can be made of dark material that absorbs and retains heat and then slowly releases it throughout the day. When the sunlight is not available, insulating blinds can be drawn to prevent heat from leaving the room.

This kind of solar energy use is passive. There are no active pumps, electrical devices, or other sources of energy required. The architect or engineer studies how the sun moves through the sky at the site. They then design a structure that uses materials in the best way to let sunlight in, absorb it, and heat that structure. Such designs can also focus on ways to keep sunlight out or cool the structure down.

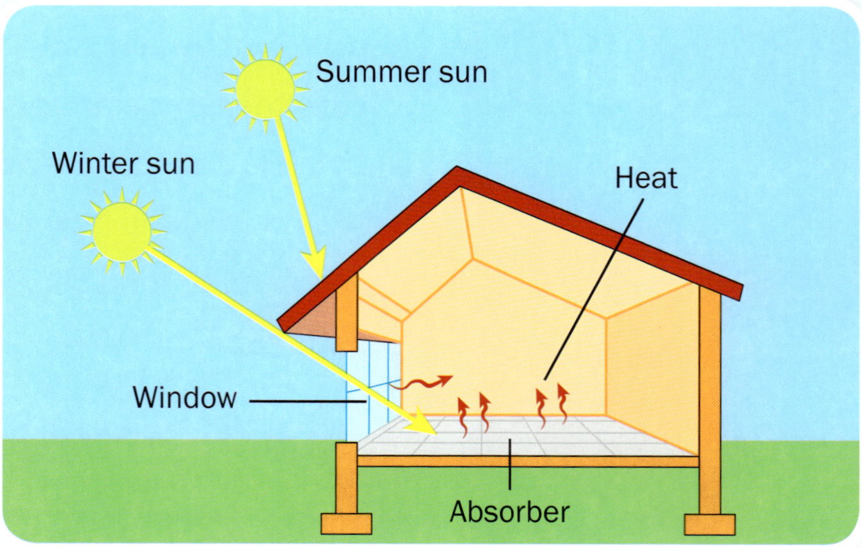


**A solar-heated home in the Northern Hemisphere will have large windows facing south and absorbent flooring that can radiate heat over the course of the day. In summer, the roof blocks the afternoon sunlight from entering the windows and providing heat that the home does not need.**

# Geothermal Energy

Chapter 8

A person purchases land to build a cabin. But there are no utility poles or pipelines nearby to carry electricity, natural gas, or water to the cabin. Instead, the builder uses solar panels to produce electricity from solar energy. For heating, they use a geothermal heat pump. The pump moves liquid through pipes that run deep into the ground and back to the home. Heat from underground, which is always about fifty degrees Fahrenheit, warms the liquid. The liquid is then pumped into a unit that collects heat from the liquid. The unit distributes water, which is now much warmer than the air outside, throughout the house.

**Big Question**

What is geothermal energy, and how do people use it?

This site in Iceland is known for its warm lake and geothermal power plant. The same thermal energy that allows people to bathe in this cool environment provides heat for homes and businesses.

This is one way of using **geothermal energy**, or the thermal energy that is underground. There are many costs and benefits of using geothermal energy.

**Vocabulary**

**geothermal energy, n.** energy produced by heat that is transferred from Earth's interior

## Requirements for Using Geothermal Energy

Another kind of geothermal energy is available when very hot parts of Earth's mantle are close to the crust. This thermal energy produces hot springs and geysers. Such heat from the mantle can only be reached in certain locations, but it can have many benefits.

This kind of geothermal energy can be harnessed to heat homes, businesses, and other structures. Geothermal heat is used to create steam. The steam or hot water is passed through pipes that heat water in a boiler. This water can pass through pipes in a home and release heat as steam. It can also be used to cook with or bathe in.

You cannot just dig a hole in the ground and find a source of steam or hot water. Most of these sources of steam or hot water are in areas where Earth's crust is active, such as volcanic areas. A well is drilled through the first layer of rock into the reservoir where steam or hot water is. After the geothermal power plant converts the thermal energy into electricity, the cooled water is pumped back into the reservoir through a different well. This recycles the water.

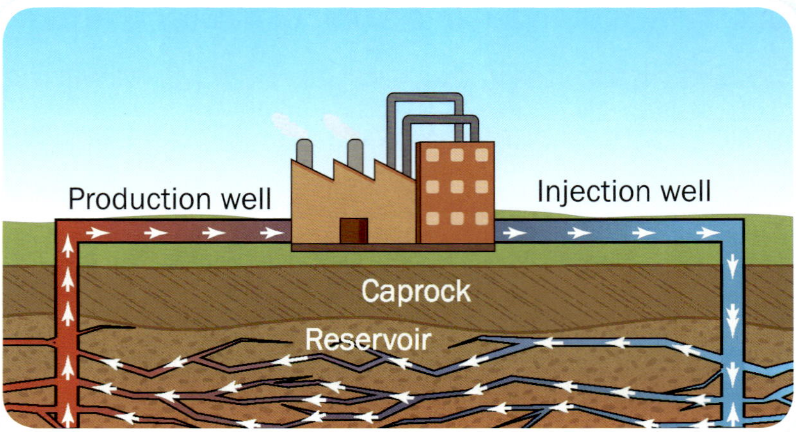


**Geothermal energy is considered clean because it produces very few greenhouse gas emissions. It also does not take anything from Earth's crust other than heat.**

## Types of Geothermal Energy Use

Another type of geothermal energy can be extracted from old oil fields. Oil fields that no longer have petroleum but are full of wastewater can be used for geothermal energy. The fluid is not clean or hot enough to directly drive a turbine. Instead, it is passed into an apparatus called a heat exchanger. This transfers the heat from the liquid to another fluid that boils at a lower temperature than water. Therefore, it turns into gas without needing to be as hot and becomes vapor to drive a turbine.

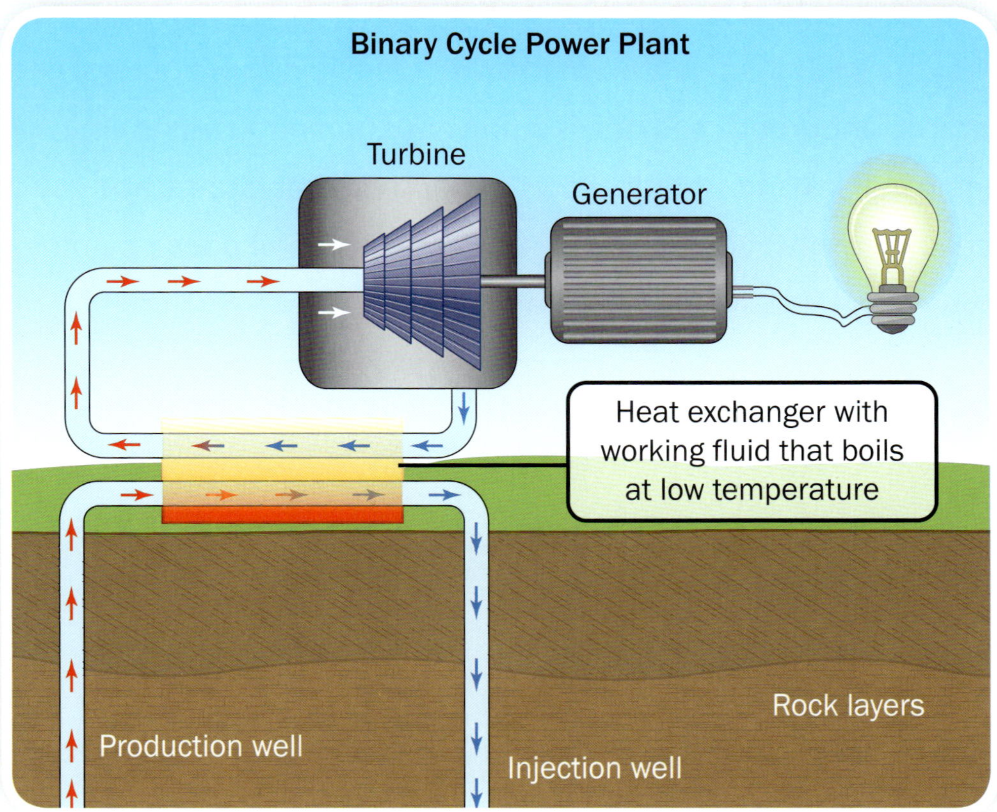


**A low-temperature geothermal power plant uses cooler fluids. The red arrows show the flow of liquids or gases that are warmer and the blue arrows show when the materials are cooler.**

## The Costs of Geothermal Energy

Geothermal energy is a clean way of meeting our energy needs. However, it cannot be done on a large scale without a big investment of money. A safe site where geothermal fluid can be tapped must also be found. Iceland and the Philippines are two countries that meet a lot of their energy needs using geothermal energy. But both have access to geothermal reservoirs thanks to their locations. There are many volcanoes in the area. The United States has far fewer volcanic areas. But it still uses geothermal energy to generate electricity in some places.

Drilling the wells for a geothermal power plant can cause natural gas to escape. Unburned methane is a powerful greenhouse gas. Any opening in Earth's crust that allows trapped methane to escape could cancel the positive effects of operating a geothermal power plant.

**Sources of U.S. Electricity Generation, 2017**

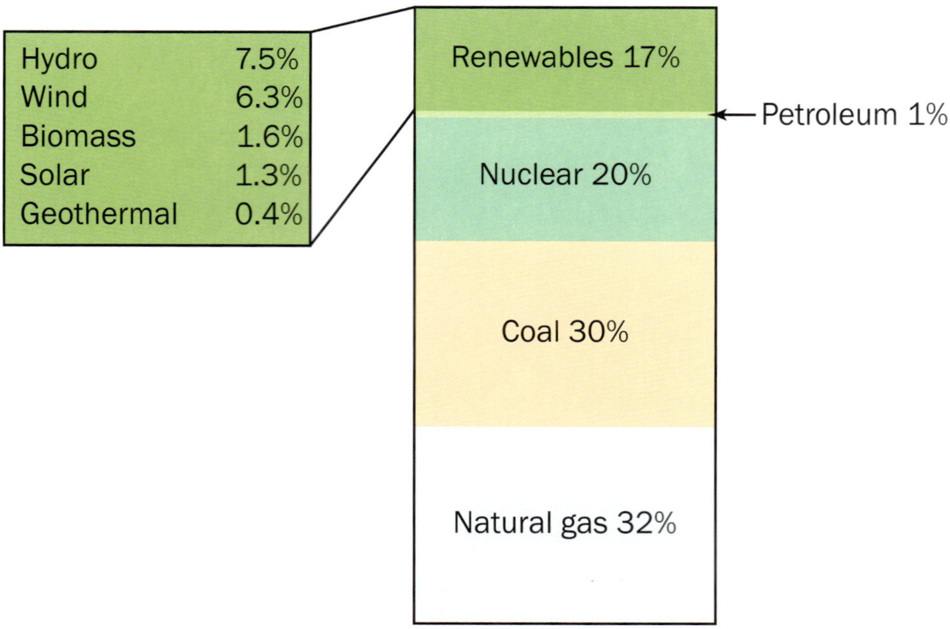


**Natural gas, coal, and nuclear power are the three biggest producers of electricity in the United States. Renewable resources account for seventeen percent in total, with geothermal providing just 0.4 percent.**

# Energy Resource Innovations

Chapter 9

You have learned about nonrenewable energy resources such as fossil fuels and nuclear power. You have also learned about renewable energy such as solar, wind, tidal, geothermal, and hydroelectric power. People are developing other resources of energy as well.

**Big Question**

What other energy resources are being developed?

The ocean floor and the layers of sediment under Arctic permafrost contain reservoirs of oil and natural gas. They also contain deposits of a fossil fuel known as methane hydrate. It is an icelike substance made of methane and water.

People do not know the potential impacts of extracting methane hydrate from the ocean floor or Arctic permafrost. More research is required to learn how to handle, transport, and use methane hydrate.

**Methane hydrate is sometimes called "fire ice" because it looks like snow but can be lit with a match.**

## People Are Developing Biofuels

**Biofuels** are made of carbon-rich chemicals that form in living things. They are produced from algae, corn plants, and sugarcane. One popular biofuel is ethanol. Ethanol is made by allowing the remains of plants, such as corn husks and stalks, to ferment, or break down in a container without oxygen. This is a type of alcohol that can be mixed with other fuels such as gasoline and burned.

**Vocabulary**

**biofuel, n.** a combustible (burnable) fuel made from the remains of recently living organisms

One benefit of using biofuels is that they are renewable. A major cost of biofuels is that they put carbon dioxide into the atmosphere and contribute to changing climate. As a result, they may not be much better for the environment than fossil fuels. They may also take up a lot of land space that could be better used for other things.

Sugarcane can be processed into ethanol, a biofuel.

## Renewable Energy for Cars

Electric automobiles and hybrid automobiles use electricity. The electric motors in these vehicles can generate electricity when the brakes are applied. The energy of motion of the car moving forward is converted to mechanical energy inside the motor. That mechanical energy then becomes electricity, which is stored in a battery. The battery supplies electricity to the motor when the driver presses the accelerator pedal to make the car move. This is called regenerative braking.

> **Vocabulary**
>
> **hybrid, n.** a combination of two types; a vehicle that uses two different sources of fuel

Cars that run on batteries either have a second engine that runs on gasoline, or they must be plugged into an electrical outlet for the battery to be charged between uses. Cars with second engines are called **hybrids**. Cars that are fully electric do not use any gasoline or diesel at all.

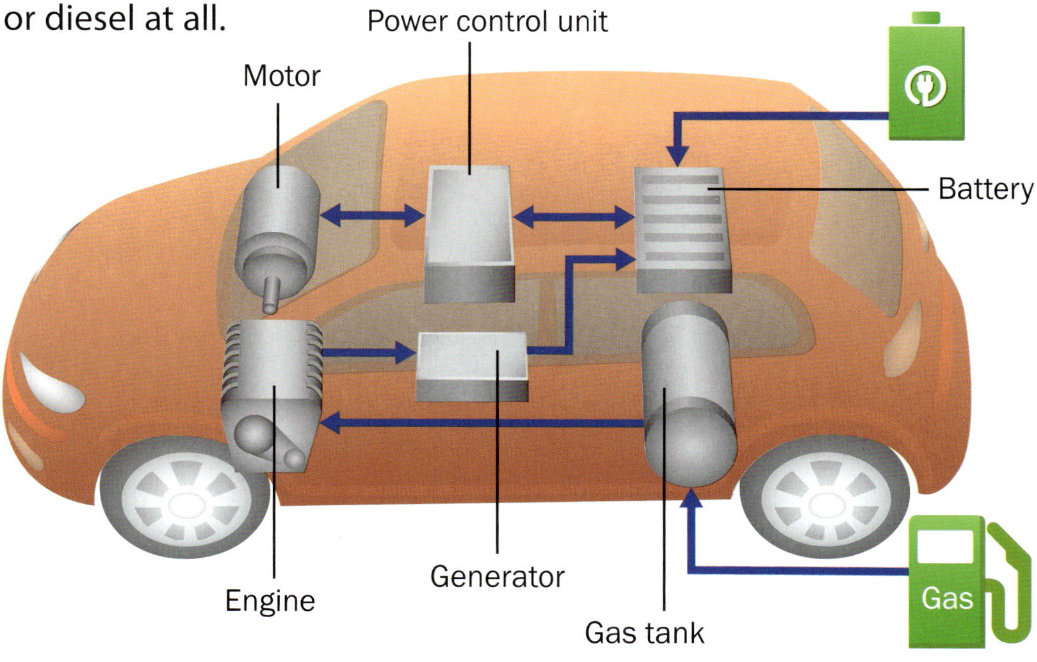


**In a hybrid car, electricity is generated by both a gasoline-powered engine and regenerative braking technology. Some can also be plugged into an electrical outlet to be charged.**

## People Are Developing Fuel Cells

Another type of fuel used in some automobiles is hydrogen. Hydrogen gas can be produced by running electricity through water. The gas is captured and compressed into a tank, which is placed in a vehicle and filled at a station. When the hydrogen reacts with oxygen in the **fuel cell**, electricity is generated. This electricity powers the motor, and the motor recovers energy while braking. It stores that energy in the battery.

Hydrogen fuel cells can be used in buses, trucks, and some smaller automobiles. The energy that is used to produce and transport the hydrogen gas is often produced by burning fossil fuels. The overall effect of using hydrogen fuel cells is not necessarily a major improvement on the use of gasoline or diesel technology. There might be less pollution coming out of the vehicles but about as much pollution going into the atmosphere overall.

**Vocabulary**

**fuel cell, n.** a device in which a chemical reaction takes place to produce electricity, similar to a battery

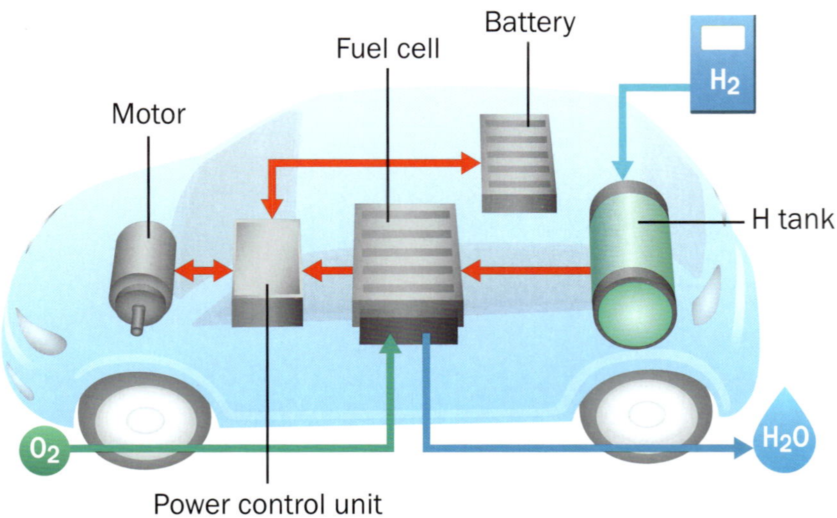


**Hydrogen gas can be pumped into a tank. The hydrogen reacts with oxygen in the fuel cell, producing electricity for the motor and the main battery. Regenerative braking helps charge the battery, too.**

## New Materials Can Harness Energy of Motion

Scientists hope to capture some of the energy that humans use and release through everyday activities such as walking and jogging. One example is a sidewalk made of triangular tiles that fit together in a puzzle-like pattern. Each corner of each tile has a small generator. When the tile is pressed downward as someone steps on it, the generators produce small amounts of electricity. Just one person walking on a few tiles does not produce much energy. But a lot of people walking on such a sidewalk all day can add up.

There is also technology being tested in clothing that involves special threads that interact as the clothing moves to produce small amounts of electricity. Will people one day plug their clothing into the grid to add electricity or reduce their electric bill? Probably not. But small-scale electricity production could be enough to charge a mobile phone or power another small device.

**Each step on this experimental sidewalk captures energy of motion and converts it into electricity.**

## Energy from Waves

Engineers can harness the energy in ocean waves through different types of technology. One example is a point absorber. It is a floating device that moves up and down with the waves. It is attached to a stem anchored to the seafloor. A piston connects the pieces. As the floating part bobs up and down, the piston generates electricity. A series of point absorbers can be placed in an area with regular wave action and then connected by a cable to a grid onshore. Because waves are made by wind, this is really another way of harnessing wind energy. In this case, water delivers the motion instead of blades of a large turbine.

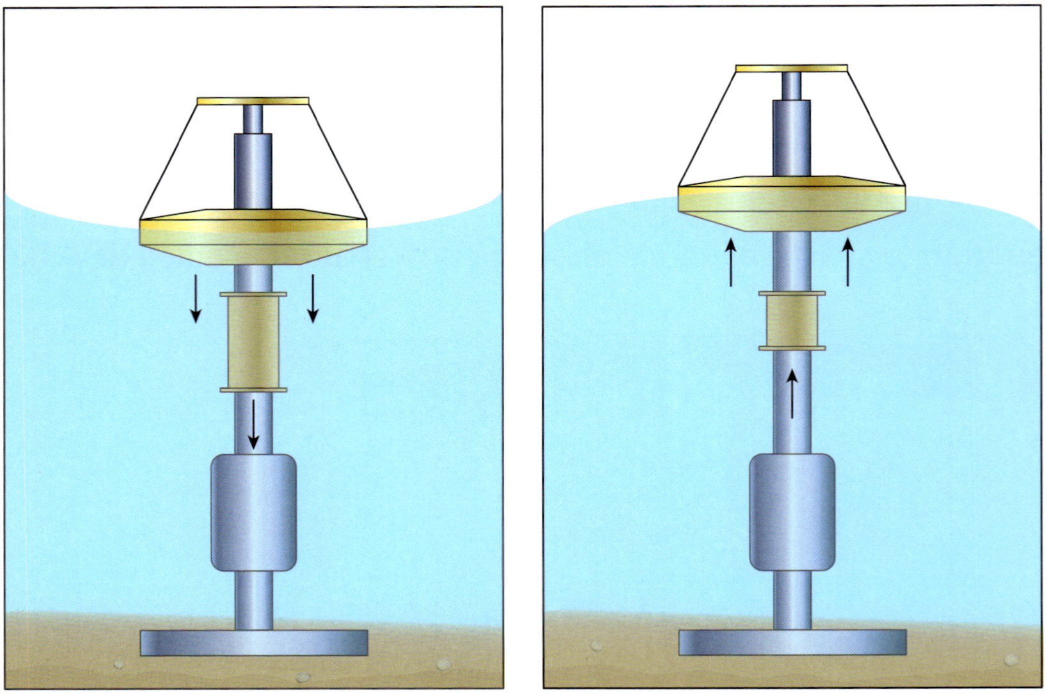

**When the low point, or trough, of a wave passes through, the point absorber drops. When the high point, or crest, passes through, the point absorber is raised. This motion transforms mechanical energy into electricity that can be sent to the shore.**

# Glossary

## B

**biofuel, n.** a combustible (burnable) fuel made from the remains of recently living organisms (48)

## C

**cost–effective, adj.** describing the balance of factors when benefits outweigh costs (33)

## E

**electricity, n.** a form of energy resulting from the flow of charged particles (3)

## F

**fossil fuel, n.** a fuel formed from the fossilized remains of organisms (2)

**fuel cell, n.** a device in which a chemical reaction takes place to produce electricity, similar to a battery (50)

## G

**geothermal energy, n.** energy produced by heat that is transferred from Earth's interior (43)

**greenhouse gas, n.** a gas that traps heat in Earth's atmosphere (20)

## H

**hybrid, n.** a combination of two types; a vehicle that uses two different sources of fuel (49)

**hydroelectric power, n.** electricity generated by the energy in moving water, usually through a dam (31)

## N

**nonrenewable resource, n.** a resource that cannot be restored as quickly as it is used (1)

**nuclear energy, n.** energy produced by the splitting of atoms, tiny particles of matter (21)

## R

**renewable resource, n.** a resource that can be restored more quickly than it is used up (1)

## S

**solar cell, n.** a device that converts sunlight to electricity (37)

**solar power, n.** the use of sunlight to meet energy needs (37)

## W

**wind farm, n.** many wind turbines installed together in an area to generate electricity for a community (28)

Core Knowledge®

CKSci™

Core Knowledge Science™

**Series Editor-in-Chief**

E.D. Hirsch Jr.

**Editorial Directors**

Daniel H. Franck and Richard B. Talbot

## Subject Matter Expert

Lisa M. Landino, PhD
Garrett-Robb-Guy Professor,
Chemistry Department
College of William and Mary
Williamsburg, VA

## Illustrations and Photo Credits

Alex Brylov / Alamy Stock Photo: 1
America / Alamy Stock Photo: 3c
Amnat Buakaew / Alamy Stock Photo: 48
andrea crisante / Alamy Stock Photo: 3b
Arterra Picture Library / Alamy Stock Photo: Cover C, 28–29
B Christopher / Alamy Stock Photo: 51
Barry Mason / Alamy Stock Photo: 9a
Cropper / Alamy Stock Photo: 3d
Design Pics Inc / Alamy Stock Photo: 25
Dorling Kindersley ltd / Alamy Stock Photo: 21
Duy Phuong Nguyen / Alamy Stock Photo: Cover B, 31
Erik Bobeldijk / Alamy Stock Photo: 30
Everett Collection Inc / Alamy Stock Photo: 18
Frantisek Staud / Alamy Stock Photo: 43
Grant Tiffen / Alamy Stock Photo: 20
Hemis / Alamy Stock Photo: 36
Jaromír Chalabala / Alamy Stock Photo: 6
Jim West / Alamy Stock Photo: 40–41
Jorn Pilon / Alamy Stock Photo: 13
kpzfoto / Alamy Stock Photo: 8
Leon Werdinger / Alamy Stock Photo: i, iii, 11
Michael Flippo / Alamy Stock Photo: 9c
Natasha Breen / Alamy Stock Photo: 7a
National Geographic Image Collection / Alamy Stock Photo: 47
olivier bourgeois / Alamy Stock Photo: 38–39
Panther Media GmbH / Alamy Stock Photo: 4
Pavel Pavalanski / Alamy Stock Photo: 32–33
Regis Martin / Stockimo / Alamy Stock Photo: 9d
Rob Carter / Alamy Stock Photo: 19
Rostislav Zatonskiy / Alamy Stock Photo: 37
Steve Taylor ARPS / Alamy Stock Photo: 3a
Sunpix Abstract / Alamy Stock Photo: 9b
trekshots / Alamy Stock Photo: 2a
US Coast Guard Photo / Alamy Stock Photo: 14
Valery Voennyy / Alamy Stock Photo: 12
Volodymyr Goinyk / Alamy Stock Photo: 2b
Volodymyr Maksymchuk / Alamy Stock Photo: 15
Westend61 / SuperStock: 24